SNAKE FARMING SECRETS UNVEILED

From Passion To Prosperity: Master The Art
Of Snake Husbandry, Breeding, And Business
For Enthusiasts And Entrepreneurs

PRESTON DAMIEN

Contents

DISCLAIMER

The information provided in this book is intended for general informational and educational purposes only. The author of this book is not engaged in rendering professional or veterinary advice. The content in this book is based on the author's personal experiences, research, and knowledge, and it is not a substitute for professional advice.

The practices and techniques described in this book are meant to provide a foundation for understanding, but specific circumstances may require individualized approaches. It is recommended that readers consult with experienced veterinarians for guidance.

The author and publisher do not guarantee the accuracy, completeness, or timeliness of the information presented in this book. The reader is responsible for ensuring that the practices and recommendations align with current laws, regulations, and safety standards in their respective locations.

By reading this book, you agree that the author will not be held responsible for any actions you take based on the information presented in this book. You are solely responsible for any consequences or outcomes resulting from your farming endeavors. Always exercise caution and seek professional advice when necessary to ensure the well-being of your farm animals and to comply with all relevant laws and regulations.

CHAPTER ONE

An Overview Of Snake Farming

Snake farming, also known as serpenticulture, is the breeding and nurturing of snakes for a variety of uses. It has grown in prominence because of the increased demand for snake-related items such as venom for medical research, snake skins for fashion, and the exotic pet trade. Snake farming is the regulated and long-term cultivation of snake species to ensure their survival while serving human demands.

The Value Of Snake Farming

Snake farming is critical to the conservation of snake species and the reduction of pressure on wild populations. It helps to reduce unlawful poaching and environmental degradation by offering an alternative supply of items such as snake venom and skins. Snake farm venom is used to make antivenom, which is essential for treating snakebite patients.

Additionally, snake farming benefits the economy by providing jobs and generating revenue from the sale of snake-derived items.

The Evolution Of Snake Farming

Snake farming dates back to ancient civilizations when snakes were adored and even grown for religious or medical purposes. When the need for snake venom for medical research increased in the early twentieth century, modern snake farming began. Snake farming evolved to encompass the fashion industry and pet commerce.

Snake Farming's Advantages And Difficulties

Snake farming has several advantages, including conservation, economic opportunity, and a steady supply of vital materials. It does, however, have hurdles, including ethical snake care, illness control, and maintaining sustainable operations that do not impact natural populations.

Striking a balance between these advantages and disadvantages is critical for the proper growth of snake farming companies across the world.

Farming Snakes Include The Following Species

Snake farming, also known as serpenticulture, includes a wide variety of snake species, each with its own set of qualities and functions. Snakes for farming may be divided into two types: venomous and non-venomous.

Snakes With Venom

Venomous snakes, while more dangerous to handle, are frequently farmed for their venom, which has a variety of medical and scientific purposes. Examples include the Eastern Diamondback Rattlesnake (Crotalus adamanteus) and the Brazilian Pit Viper (Botbothrops jararaca). Venom extraction is a sensitive procedure in which venom is carefully collected and processed for the creation of antivenom, pharmaceutical research, or even as a component in cancer research.

Snakes That Are Not Venomous

Non-venomous snakes are less dangerous to handle and are usually raised for their skins, meat, or as unusual pets. The Ball Python (Python regius) and the Corn Snake (Pantherophis guttatus) are two popular non-venomous snake species for farming. These snakes are quite placid and available in a variety of eye-catching patterns and colors, making them popular as pets and fashion items.

Farmable Snake Species Include

Some of the numerous snake species cultivated have attracted special attention because of their economic and ecological value. The Burmese Python (Python bivittatus), a Southeast Asian native, is widely bred for its gorgeous skin, which is utilized in the fashion business. Furthermore, the Texas Rat Snake (Pantherophis obsoletus) is bred on farms as an effective rodent control agent, assisting in the natural management of pest populations.

In summary, snake farming comprises a diverse spectrum of snake species, including both venomous and non-venomous snakes, with each having a distinct function, such as venom extraction, skin and meat production, and pet trade. Specific species' popularity varies according to their utility and demand in various sectors.

Creating A Snake Farm

Creating a snake farm involves careful planning. First, choose an ideal location by taking into account issues like accessibility, closeness to veterinary facilities, and local legislation. Create a plan that allows caregivers to move freely, reduces stress on snakes, and incorporates quarantine rooms for new arrivals. Waste disposal systems and quarantine facilities must be included in the infrastructure. Make biosecurity measures a top priority to avoid disease outbreaks.

Choosing An Appropriate Site

Choosing an appropriate site for a snake farm is critical. Choose a location with a climate suited for the target snake species. Make sure there is enough room for expansion and that zoning restrictions are followed.

Maintaining optimum humidity levels is aided by proximity to a water supply. Conduct a thorough risk assessment, taking into account issues such as possible floods and other environmental threats. A well-chosen location is critical to the profitability and long-term viability of the snake farming operation.

Snake Enclosures And Habitats

Building proper enclosures is critical for snake health and well-being. Create cages that are similar to their native environments, with hiding locations, climbing structures, and appropriate substrate. Enclosures should be escape-proof and easy to clean.

Assess and change the environment regularly based on the snake's behavior and growth.

Temperature And Humidity Control

Snake health must maintain ideal temperature and humidity levels. Install a dependable climate control system to mimic the snake's natural habitat. Monitor and change temperature and humidity factors regularly, taking into account the unique demands of each snake species. Failure to control these elements might result in stress, disease, or even death.

Safety Precautions And Rules

In snake farming, safety precautions and rules are non-negotiable. Staff should be trained on proper handling practices and first aid.

Create emergency reaction strategies in the event of an escape or an accident. Comply with municipal, state, and federal rules governing snake farming, such as permit requirements and record-keeping.

Establish health regimens and illness prevention methods in collaboration with veterinary specialists. A dedication to safety and compliance not only protects human welfare but also maintains the snake farming enterprise's ethical and legal status.

CHAPTER TWO

Snake Farming Feeding And Nutrition

Feeding and nutrition are important factors of snake farming success. Snakes are carnivorous reptiles with specialized nutritional requirements. Proper diet is critical to their health and growth. Snakes eat mostly rodents, birds, and other small creatures. To suit their dietary requirements, the nutritional composition of their prey must be balanced. Obesity, vitamin inadequacies, and poor calcium-to-phosphorus ratios are all common dietary issues in confined snakes. Snake growers should provide a range of prey items and consider supplementing their meals as needed with vitamins and minerals.

Snake Diet And Feeding Routine

A constant feeding schedule is critical in snake farming. Most snakes do not require frequent feedings and instead have erratic feeding habits.

Smaller snakes may consume more food than larger ones. It is critical to carefully match their normal eating patterns as much as possible. Obesity can be caused by overeating while undereating can hinder growth. The feeding plan should be tailored to the snakes' species, age, and individual requirements.

Frozen Vs. Live Prey

One of the controversies in snake farming is whether to use live or frozen prey. Although live prey might enhance a snake's hunting instincts, it can also bring concerns such as snake damage. Frozen prey, on the other hand, is more secure and easier to preserve.

Many snake farmers prefer to use pre-killed frozen rats since they reduce the risk to the snake and make feeding easier. Some snakes, however, are picky feeders and prefer live prey. The choice of live or frozen prey is determined by the snake's species, age, and personal preferences.

Managing Feeding Issues

Feeding issues in snake farming might include snakes refusing to eat or regurgitating their meals. These problems can be caused by stress, poor enclosure conditions, disease, or the wrong prey size. Taking on these issues demands perseverance and keen observation. Snake farmers should provide a stress-free habitat, maintain correct cage conditions, and give prey of sufficient size.

Consultation with a veterinarian or expert snake breeder can also assist with feeding problems. To preserve the snakes' health and well-being in captivity, it is critical to carefully monitor them and alter their nutrition and care as needed.

Snake Farming Breeding And Reproduction

Breeding and reproduction are critical parts of snake farming, as they are required for sustaining captive populations and developing desired snake features. Successful snake breeding necessitates a thorough study of the species at hand. Most of the time, it entails carefully selecting breeding pairings based on genetics, health, and temperament. Because of their manageable size and docile temperament, ball pythons and corn snakes are attractive alternatives for breeding.

Recognizing Snake Reproduction

Snake reproduction is mostly sexual, with male snakes transmitting sperm to female snakes via specialized organs known as hemipenes. Understanding snake reproductive anatomy and behavior is critical for breeding success. Before copulation, snakes frequently participate in intricate courting rituals that differ by species.

Seasons And Cycles Of Breeding

Many snake species have distinct mating seasons and cycles. Temperature, daylight duration, and humidity can all have an impact on these. Ball pythons, for example, frequently breed in the cooler months, but corn snakes breed in the spring. To increase reproductive behavior, snake farmers must mimic similar settings in captivity, utilizing controlled lighting and temperature regimes.

Taking Care Of Snake Hatchlings

Snake hatchlings require particular attention after they emerge. Hatchlings are vulnerable to stress, dehydration, and inappropriate nourishment. It is critical to have proper housing, temperature, and humidity conditions.

Snake farmers must also choose suitable food, such as tiny rodents or insects, and ensure that hatchlings are fed regularly.

To summarize, successful snake farming requires a thorough grasp of snake reproduction, rigorous management of breeding seasons, and meticulous care for hatchlings to produce healthy and thriving snakes for the pet trade or conservation initiatives.

CHAPTER THREE

Health And Disease Management

To maintain the well-being of the reptiles and the profitability of the farming operation, health, and disease management are critical in snake farming. illness prevention is frequently more successful than illness treatment. Adequate husbandry techniques, such as maintaining adequate temperature and humidity levels, giving clean water, and providing an appropriate feed, are critical for snake health. Regular health checkups performed by qualified individuals can identify early indicators of sickness.

Snakes With Common Health Problems

Snakes may have lung illnesses, skin infections, mites, and parasites, among other things. Incorrect temperature or humidity conditions in their cages can cause respiratory illnesses.

Skin infections can arise as a result of trauma or poor environmental conditions.

Mites and parasites can hurt snake health and must be treated as soon as possible.

Preventive Medicine

Preventative maintenance entails keeping ideal circumstances for snakes to thrive. Proper enclosure design, temperature regulation, and cleanliness are all part of this.

Cleaning cages regularly, giving clean water, and providing balanced food with the required vitamins will help to prevent many health conditions.

Identifying And Treating Snake Diseases

Snake farmers should be taught how to spot indicators of sickness, such as changes in behavior, hunger, or skin condition. When an illness is detected, quick intervention is critical. Isolation of ill snakes, consultation with a reptile veterinarian, and adherence to prescribed therapies are all necessary procedures.

Snake Veterinary Care

Experienced reptile doctors are essential in snake farming. They can diagnose and treat complicated health problems, provide routine check-ups, and advise on preventative measures. Regular veterinarian checkups are recommended to preserve the snake farm's general health and output. Snake farmers should develop an excellent working connection with a reptile veterinarian to ensure the health and prosperity of their reptiles.

Extraction Of Venom:

Venom extraction is a vital step in snake farming, to obtain venom from dangerous snakes for research, antivenom manufacturing, and pharmaceutical uses. It entails securely collecting venom without injuring the snake. This approach not only advances science but also assists in the conservation of these reptiles.

The Function Of Venom Extraction:

In the worlds of herpetology and medicine, venom extraction plays various vital functions. For starters, it ensures a consistent supply of venom for studies on venom composition, prospective medicinal uses, and evolutionary biology. Secondly, snake venom is necessary for the production of antivenin, which is essential for the treatment of patients bitten by snakes. Furthermore, venom extraction can assist in reducing human-snake conflicts by offering a safer method of collecting venom, lessening the temptation to kill snakes.

Venom Extraction Tools & Techniques:

To protect the safety of both the snake and the handler, venom extraction requires specific gear and methods. Snake hooks, restraining tubes, and venom-collecting containers are common instruments. Handlers use these instruments to gently hold the snake's head, enabling venom to be safely extracted from its fangs via a technique known as "milking." Milking includes applying mild pressure to the fangs, allowing the venom to flow into a collecting vessel.

Venom Handler Safety Protocols:

To avoid mishaps and snakebites, venom handlers must follow stringent safety standards. To reduce the chance of venom exposure, they use suitable protective gear like gloves and eye protection. Handlers are trained on snake behavior and safe handling practices to ensure that the snakes are not stressed. Immediate medical assistance is required in the event of a snakebite.

In conclusion, venom extraction is critical in snake farming, since it supports research, antivenom manufacture, and conservation initiatives. It uses specific instruments, procedures, and safety regulations to safeguard both the snakes and their handlers, making it an important practice in herpetology and medicine.

Snake Farming For Monetary Gain

In recent years, snake farming, also known as serpenticulture, has gained traction as a viable business endeavor. It entails the breeding and raising of several snake species for commercial purposes. The major goal is to meet the growing demand for snake-related items such as venom extraction, leather, and meat. Snake farming has various advantages, including endangered species protection, sustainable sourcing, and economic prospects.

Demand And Market Analysis

For success in the snake farming sector, a detailed market study is required. Pharmaceutical enterprises seeking venom for antivenom manufacture, fashion industries employing snake leather, and unusual meat markets are all interested in snake-related items.

Cultural preferences, medical uses, and international trade rules all have an impact on market dynamics. Understanding these elements is critical for long-term growth and profitability.

Snake Goods And Byproducts Include

Snake farming produces a variety of items and byproducts. Snake venom is a vital resource for the creation of antivenom, and snake leather is sought after for luxury fashion products.

Snake flesh is also considered a delicacy in some areas. Bones and scales are byproducts that are used in traditional medicine and ornamental arts. Maximizing the use of these resources is critical for a thriving snake farming business.

Distribution And Marketing

In the snake farming sector, effective marketing and distribution tactics are critical. An effective marketing strategy includes targeting specialized markets, cooperating with pharmaceutical firms, and creating ethical and sustainable procedures. Efficient distribution channels guarantee that snake-related items, whether local, national, or worldwide, reach their target consumers.

CHAPTER FOUR

Conservation And Ethical Considerations

Snake farming has a dual purpose in terms of conservation and ethical issues. On the one side, technology can help conservation efforts, but it also poses ethical concerns that must be addressed.

Snake farming, from a conservation standpoint, can assist in alleviating pressure on wild snake populations by providing a sustainable source of snakes for the pet trade, traditional medicine, and research. It is possible to prevent unlawful poaching and environmental degradation by supplying the demand for snakes through cultivation.

However, ethical issues arise when examining the conditions under which snakes are housed and the care they get.

The Conservation Importance Of Snake Farming

Snake farming can help conservation by reducing reliance on wild populations for a variety of uses. Snake farms can help maintain biodiversity by reducing the over-exploitation of wild snake species through controlled breeding and sustainable procedures. Furthermore, snake farming frequently gives possibilities for research and teaching, which helps us understand these species and their environments.

Ethical Practices

Ethical measures in snake farming are critical to ensuring the animals' well-being. This involves providing appropriate cages, correct feeding, veterinary treatment, and stress-reducing handling procedures. Breeding systems that promote genetic variety and reduce inbreeding are also ethical issues. Transparency in sourcing and compliance with animal welfare regulations are critical in ethical snake farming.

Compliance With The Law And Regulations

To operate ethically and sustainably, snake farming must abide by applicable rules and regulations. Compliance includes getting the proper permissions, adhering to humane treatment norms, and engaging in conservation initiatives as appropriate. Regular inspections and audits are critical to ensuring that farms comply with the law, preventing concerns such as illicit wildlife trading and exploitation.

Finally, when handled responsibly and by legal requirements, snake farming may positively contribute to conservation efforts. Addressing ethical concerns and maintaining regulatory compliance is critical for the appropriate operation of snake farming.

Future Snake Farming Trends

The future of snake farming is set to undergo dramatic changes as a result of changing consumer demands and conservation activities. Snake farming is set to expand outside conventional markets as worldwide demand for unusual foods develops. A hopeful trend is the introduction of specialized breeding programs for uncommon and endangered snake species, which aligns snake farming with conservation aims. Furthermore, as snake venom research becomes increasingly important for medicinal developments, there is a possible spike in pharmaceutical sector interest. The use of genetic technology for selective breeding may potentially result in the generation of snakes with desired characteristics.

Advances In Snake Farming Technology

Technology is transforming snake farming, increasing efficiency and safety. From temperature control to feeding schedules, automated monitoring systems coupled with sensors and AI offer ideal circumstances for snakes. Genetic testing aids in the discovery of desirable features and aids in the preservation of genetic variety. Drones and robotics automate duties like enclosure cleaning and snake handling, minimizing the need for humans to contact with animals. Snake handlers may practice their skills with virtual reality and augmented reality software. Farmers can monitor snake health in real-time using precision farming techniques such as data analytics and IoT devices. These advances not only increase output but also benefit captive snakes.

Sustainable Farming Practices

Snake farming is adopting more sustainable ways as environmental awareness rises. This includes eco-friendly cage designs that resemble natural environments, lessening farming's environmental effect.

Sustainable feeding strategies include using ethically sourced prey and reducing waste. Controlled breeding programs attempt to conserve species by avoiding over-exploitation of wild snake populations. Renewable energy sources, such as solar electricity, are increasingly being integrated into snake farm operations, helping to reduce the carbon impact. Partnerships with local communities for habitat protection and education also help to ensure the long-term viability of snake farming. These techniques not only solve environmental issues, but they also appeal to consumers who value ethical and ecologically friendly products.

Conclusion On Snake Farming

To summarize, snake farming is a complicated and contentious business that has sparked both curiosity and worry. As demand for snake-based items such as meat, skin, and venom grows, it is critical to achieve a balance between economic and ethical issues.

Snake farming may help local economies while also reducing pressure on wild snake populations, but it must be done properly and sustainably. To preserve the well-being of these species, limit the spread of zoonotic illnesses, and conserve ecosystems, strict rules and ethical principles should be in place.

Furthermore, public education and knowledge about the value of snakes in preserving ecological balance, as well as their function in medicine and research, are critical. Governments, conservationists, and the business itself must work together to guarantee that these methods support both human livelihoods and the preservation of our natural world.

Finally, the success of snake farming is dependent on our capacity to make educated judgments that balance human interests with the well-being of these sometimes misunderstood species.

www.ingramcontent.com/pod-product-compliance
Lightning Source LLC
Chambersburg PA
CBHW060014300526
45794CB00003B/1188